Google Classroom

A Simple, Concise & Complete Guide to Take Your Classroom Digital

Table of Contents

Introduction 4

Chapter 1- How to Set up Google Classrooms 17

Chapter 2- Differentiating Assignments in Google Classroom 30

Chapter 3- Best Practices Of Google Classroom Teachers Should Know 38

Chapter 4- Great Apps to Use Together with Google Classroom 47

Chapter 5- Creative Ways to Use Google Classrooms to Teach 76

Chapter 6- 10 Things Students Can Do Using Google Classroom 90

Chapter 7- Using Google Classroom for Professional Development 97

Introduction

Congratulations on downloading your personal copy of Google Classroom (Technical) *A Simple, Concise & Complete Guide to Take Your Classroom Digital*

Google Classroom is a fantastic free-to-use application that allows for online teacher-student collaboration. For teachers, Google Classrooms allows them to create an online classroom environment and invite students to these classes. It also allows them to set up and distribute assignments.

For students, they would be able to converse with the teachers one-on-one about their assignments and classes and teachers would also be able to track a student's progress. To use Google Classrooms, schools must register for a free account at Google Apps for Education.

Google Classroom aims for a paperless study environment and experiment. This application is also used in tandem with other Google applications such as Google Docs, Google Sheets, Google Calendar, Gmail and Google Drive.

Developed in 2014, Google Classroom is part of G Suite for Education.

When it went live in 2014, it was released as a Classroom API which allowed school administrators and developers to engage in an online environment. After that, Google Classroom was updated even further to integrate Google Calendar particularly for due date requirements for assignments, class activities, field trips and so on. Come March 2017; Google opened the Classroom to allow personal Google users to join classes without having a unique G Suite ID. Finally, in April 2017, individual Google users also had the ability to create classes and teach them via Classroom.

In the next chapters, you will be given a step-by-step introduction on how to get started with Google Classroom- it is really that easy and simple so don't worry. Next, we will also look into how you as an educator can use Google Classroom to its fullest by employing best practices. You will also be introduced to an array of wonderful apps that can be used with Google Classroom.

Lastly, we will also explore the various ways to teach core subjects to students and the links that you can use as well as how Google Classroom can be use for professional development.

But before we go on, here are some interesting pieces of information to convince you on the wonderful world of Google Classroom.

Advantages & Disadvantages of Google Classroom

As with any new technology, some components and elements work and don't work or rather there could be some benefits and some issues that need to be ironed out, but that is why there are regular updates on apps. If you are thinking about using Google Classroom, here are some benefits to the app and some characteristics that could be a disadvantage.

The Advantages of Google Classrooms

1- Accessible from various devices

    Google Classroom is an incredibly easy piece of the app to use, as with all other Google products. It can be accessed through Google Chrome, which means that this app can be utilized on your computers, tablets and mobile phones anywhere, anytime as long as you have a consistent internet connection.

2- Easy to use
You can add as many learners as possible, and you can easily create Google documents that can be used to make announcements and assignments. You can also use it to post up a variety of links and even YouTube videos to enhance learning and also attach files directly from Google Drive. Students accessing the Classroom can log in and log off quickly as well as receive assignments and turn in completed ones as easy as can be.

3- Handy sharing tool
Incorporating the use of Google Docs has brought immense benefits for Google Classroom. Using Google Docs enables users to save their documents online and allow it to be shared with so many people. Each time the teacher creates an announcement or publishes an assignment via Google Doc, students will be able to access these files immediately by logging to their Classroom or Google Drive.

4- Effective Communication tool
The use of Google Docs also makes Classroom an effective communication means between students and teachers. Teachers do not need to rely on emails to share information

especially when they need to connect and communicate with a large number of students. By creating a document on Google Docs, you share to as many students as you want and, you also can organize these documents and personalize your Drive folders.

5- Fewer Excuses
Since students are connected to the internet from all their devices, especially their mobile phones, there is no way that they can miss out on working on their assignments especially if the assignments are things that need to take pictures. As a teacher, you would probably hear less of the 'my dog at my homework' problem.

6- Easily make and distribute assignments
One of the crucial factors of the success of the Google Classroom is the ability if gives teachers to create assignments online and distribute it fast with just a few clicks of their mouse. Students can turn in completed assignments as soon as they are done with, without waiting for the next school day to come around. Setting up, sending and finishing up an assignment has never been as efficient and quick as this. Google Classroom

also enables the teacher to check on assignment submission and also to track which students are still working on it. Teachers can also provide immediate feedback on these assignments.

7- Prompt feedback
   Teachers can offer online support to students promptly and immediately. This allows for feedback to be more efficient because immediate feedback will have a better impact on student's minds, as opposed to delayed feedback, when the student itself may have to remember what questions and what answers they provided for the assignment.

8- There is no need for paper
   If one of your teaching agendas is to reduce paper and also lessen the fuss around printing and distributing paper assignments to students, then Google Classroom is for you. Google Classroom utilizes the eLearning system in one all powerful cloud-based location, so it gives, teachers the ability to go paperless and eliminate the need for printing, handing out assignments, grading them once everyone has completed it and also reduces the part about the dog eating the homework.

9- User Friendly

Google Classroom employs the clean and easy to use user interface you see on all Google apps. Single design details enable users to understand and grasp how to use the app much easily and also enables the user to adapt to the app faster. The simple and intuitive interface can convert even the most least IT savvy of users.

10- Commenting system is a plus

To make online discussion even more conducive, Google Classroom also enables you to create URLs for interesting comments, enabling users to use the link to further these discussions. What's more, students can also comment on specific locations within certain assignments.

11- Use for professional development

Although this product is catered towards school classrooms all over the world, in truth this app can also be used for professional development to manage projects, send out project updates, create questionnaires and so on. It can also be used for faculty meetings as well as a group of educators.

12- Notifications

Both students and teachers can enable mobile notifications from Classroom. You can set notifications for your computer, iOS devices as well as Android devices and ways to set notifications varies from the operating system used. Essentially, teachers and students receive notifications on separate things:

Teachers	Students
• You receive an assignment the second time the student uploads it after reviewing and returning • You receive a private note from a student • An assignment is sent after the due date • Another teacher invites you to teach a class as an additional teacher. • A student comments on a	• You get invited to a class by a teacher • A new assignment, question, or announcement is created by a teacher • You receive a returned work or when you are graded • When someone comments on your post or mentions you in a post or comment. • When a teacher

post • A student mentions you inn a post or even a comment • When a scheduled post is published or failed to post. Teachers can specify which notifications you want to receive or just leave it by default.	sends you a private note. • You have unsubmitted work that's due within 24 hours. Students can specify which notifications you want to receive or just leave it by default.

There are plenty of benefits of using Google Classroom. The idea is to start using it first to see how it helps in your objective or your school's objective of steering towards a digital classroom.

Google Classroom is continuously updated with improvement and enhances but the first thing to do is to start with the basics.

The Disadvantages of Google Classroom

Depending on how you look at it, these disadvantages could also be additional points. You be the judge.

1- Troublesome account management

Google Classroom can only be accessed through one domain and not through multiple domains. You cannot use your personal Gmail to log into Google Classroom either. You need an exclusive log in ID created for the use of Google Apps for Education. This means you need probably have by now, two Google IDs- one for your personal Gmail use and the Education suite. This can make sharing items from your personal account to your Classroom troublesome as you would need to save the item to your desktop or a specific folder and then upload it to the Classroom's Google Drive.

However, the good thing is you have your personal stuff separate and your professional things in one drive. If you are a person who prefers to keep a professional account for your work and personal account for everything else, then signing up with a G Suite for Education ID would be useful. This can prevent you from sharing things you don't want to by accident, and it also enables you to separate which emails go into your personal account and work stuff goes into your professional Gmail account.

2- Requires converting formats

One hassle with using Google products is that you can only use formats that work with the Google Apps. For example, while you can upload a Word document to Google drive, you need to convert it to a Google Doc to edit the content.

3- Limited Sharing capabilities

This feature is often a good thing when you think about privacy and content protection. Google Classroom doesn't allow users to share documents if they are not owners of the material. It can be a hassle if you need to share with more than 50 users, but then again, the document owner would be allowed to share to a limitless number of people. Only those they have shared it too, cannot share the content with their peers.

4- Editing issues

When content is shared to students, students also become owners of the document, and they would be able to edit it which means certain parts of the assignment can be deleted if it happens accidentally. Therefore, assignments need to be made in PDF formats that answers can be written into.

5- No automated tests or quizzes

Google Classrooms does not have automated quizzes and tests. Therefore it does not have the ability to fully replace whatever Learning Management system you use.

6- Impersonal

Let's face it. Any learning environment online cannot replace the magic behind face-to-face interpersonal communication. Google Classrooms must be viewed as an additional to interactive educational methods that involve building friendships and working relationships in the classroom for effective interaction and to develop communication skills.

Conclusion

Google Classrooms, like all other Google products, are continuously updated to enhance user experience. So whatever elements you find a hassle now, will be rectified as time goes by.

Now that you have a good idea of the benefits and probable disadvantages of Google Classrooms, it is time to delve deeper on how you can use Google Classrooms to expand and convey active learning methods for your students.

Who Can Benefit from Google Classroom?

Basically all sorts of educators and all learners benefit from Google Classroom. The idea of Google Classroom is to bring quality education to every child or student in the world.

Google Classroom benefits those who are being home-schooled, benefits children with special needs, benefits countries that do not have access to quality resources and information and it benefits anyone looking to gain knowledge learning from an expert in the industry or subject.

The important thing to remember is that internet connection is vital because without it, you would not be able to access Google Classroom.

Chapter 1- How to Set up Google Classrooms

In order to access Google Classrooms, you first need to create an account with G Suite for Education. However, with the 2017 update, you now do not need to have a specific Suite for Education ID. All you need is Gmail account. The probable reason for allowing only a Gmail account ID is that more people outside the education system are finding it beneficial to use Google Classroom such as project managers, lecturers, office workers and so on.

But in this chapter, we will look into signing in using a G Suite for Education account.

Here is how you do it:

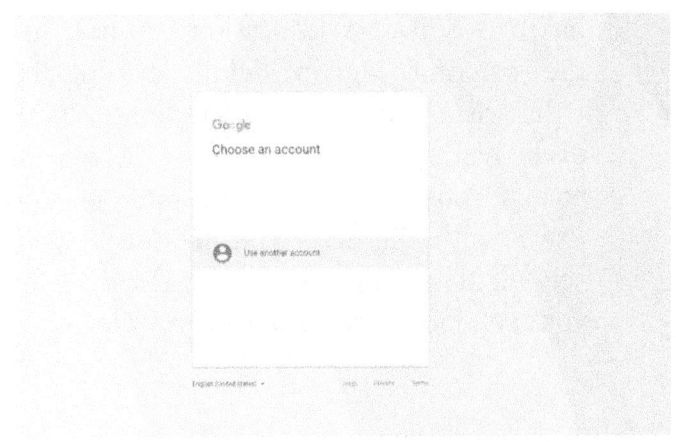

1- Open your web browser and type in classroom.google.com
2- Click enter. Google will request that you sign in using a Gmail account.

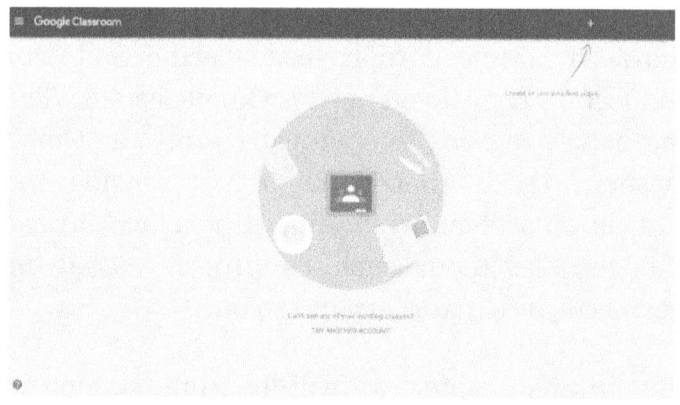

3- Click on 'Create or Join your first Class'. When you click on it, you will see this notification. If your school or university has a G Suite for Education account, use the existing log in details to log in. Otherwise, you need create an account exclusively for the suite to use it. If your sole purpose for Google Classroom is to use it with the class you are teaching in school, then best to use the ID given by your school, or create one on your own.

GOOGLE CLASSROOM

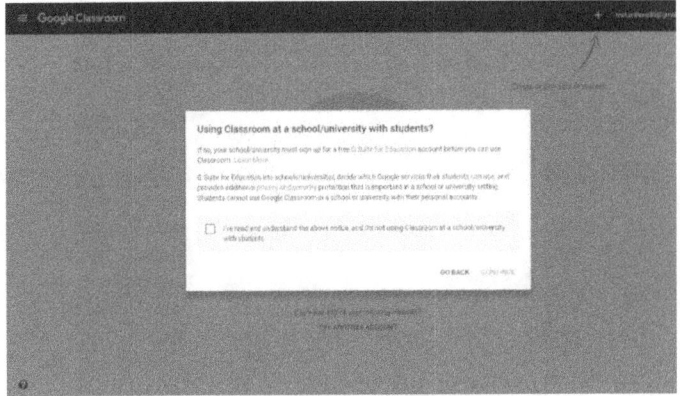

4- Click on G Suite for Education and insert the necessary details.

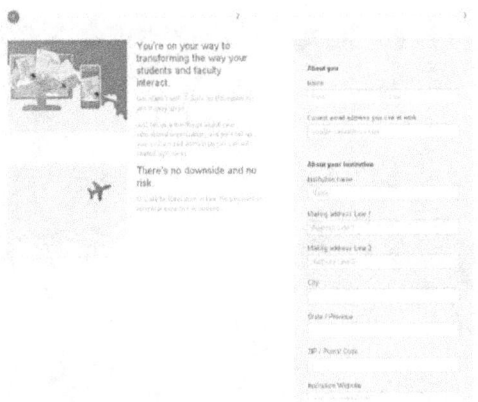

5- Once you are done logging in, you will then come to the Welcome Screen. Click the plus sign at the top right corner and choose Create Class.

6- In the Create a Class dialogue box, type in the Class Name and Section. Next, click on Create to create your class.

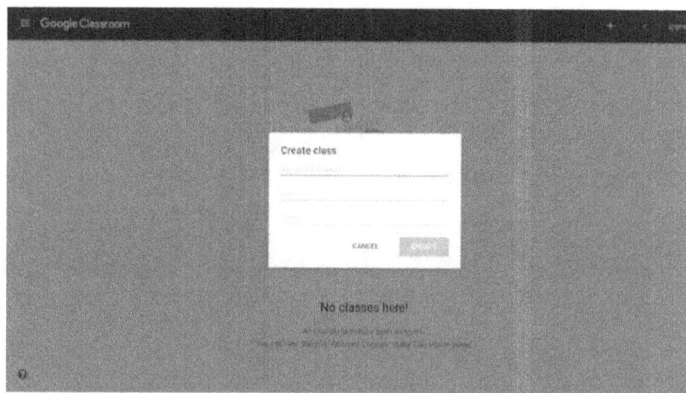

7- Your new classroom is created. In this section, you can see that there are three main tabs. The Stream, Students as well as the About tab.

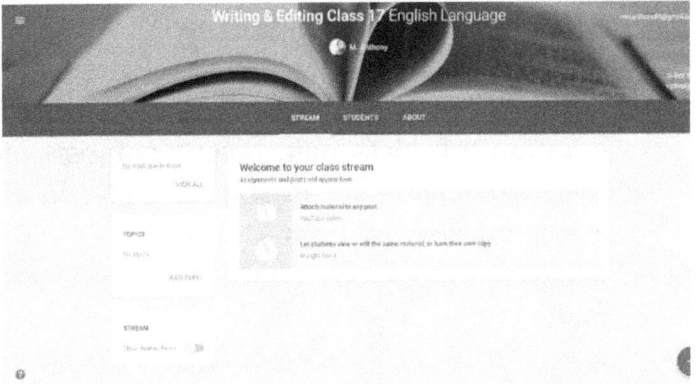

- **Stream:** This tab allows you to manage all class assignments and publish announcements to the entire class. You can add new assignments complete with deadlines and attach any other relevant materials. Upcoming assignments are shown at the left. You can also use select social media services to send messages to your entire class, even with an attachment in them.
- **Students:** This tab allows you to manage your entire class. You can invite students to your classroom from this tab and also manage their permissions level. To invite students to your class, all you have to do is add them into your Google Contacts in your Google Apps for Education account. They can also already be in the schools' directory for this purpose.

- **About:** Give the students a brief of what the course is about and what can they expect to learn. You can add in any details such as duration of class, outcomes, number of assignments and so on. You can also add in a location for the class and update it with materials for your class's Google Drive folder.

8- Upon creating your first class, you will no longer see the Welcome screen each time you log in. Instead, what you will see home screen and the thumbnails show you the classes that you have created. To access it, you just need to click on the thumbnail.

From here onwards, you can do several other things:

- **Add a new class:** You can still see the plus sign at the upper right of the screen and you can add in as many classes as you like.
- **Rename or archive a class:** Click the three dots stacked on your current class and this feature will allow you to edit your class or archive it. Archiving a class removes it from being active but still accessible. Your students can still see the contents of the class in the Google Drive but you, as the owner of the class, will not be able to make any changes to the class. You can always restore your archived classes anytime by going back to Archived Classes and clicking the three dots again and restoring it.
- **Access Google Drive for the class:** By clicking on the file icon at the bottom right corner of your Class thumbail, you would be able to access any files in your Google Drive account linked to the Classroom.

Using Google Classroom in everyday class

Once you are done covering the basics, here are some more detailed functions that you can do and use:

Adding Announcements

9- Communicate with your class by adding an announcement. Click on the Plus sign at the bottom right hand corner to display the variety of options.

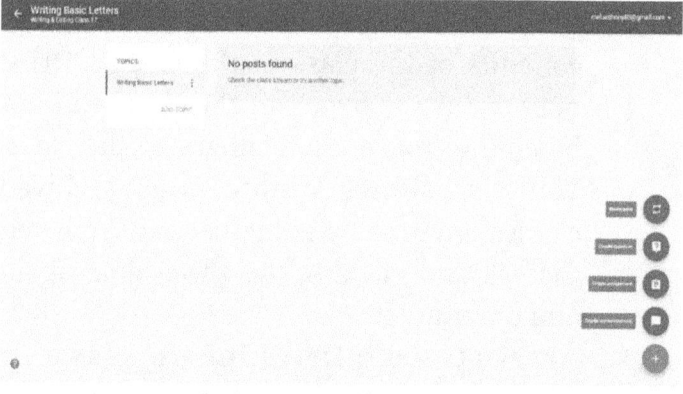

10- Click on 'Create Announcement'. A pop up box will be shown. Add in your message and attach social media links or files from your Google Drive. Once you are happy with you message, you can click Post or else click on the arrow for more options such as Save Draft or Schedule.

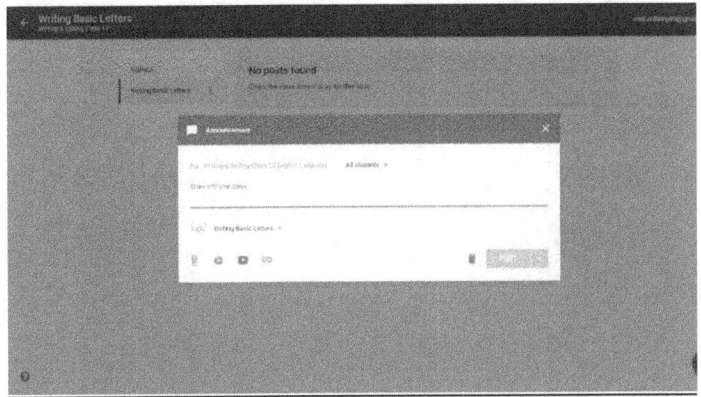

Adding Assignments

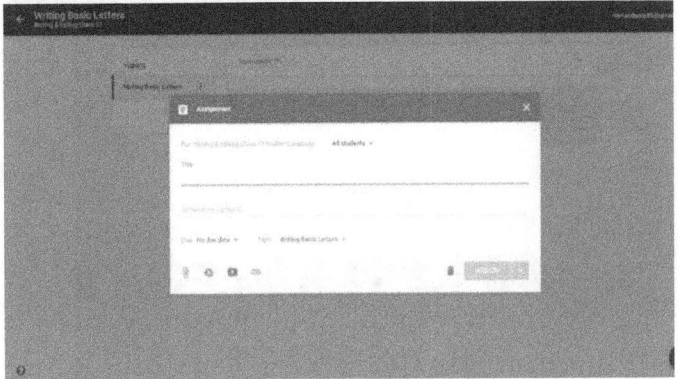

11- Adding a new assignment can be done by clicking on the same plus sign button. You can determine if the assignment has a due date as well as select which topic the assignment is under. When a student receives this assignment, there would be extra notifications on the assignment that reminds the students when their assignments are due.

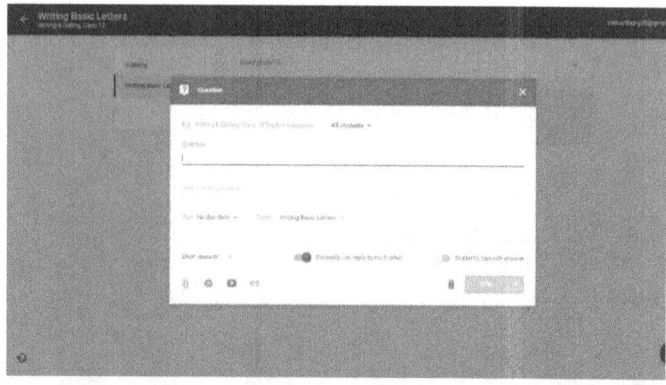

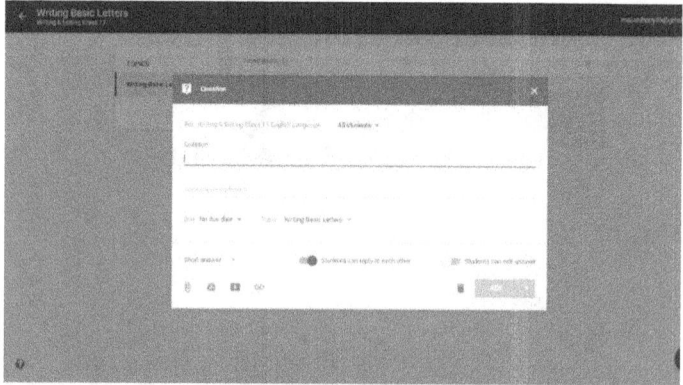

Teachers are also able to post questions and also reuse a previous post or reuse a template that works well with the current assignment they have planned. This cuts down the time and effort the teacher needs to spend on in order to create a new assignment in the same module or with the same needs.

What Else can you do in Google Classroom?
Manage details of students in class.

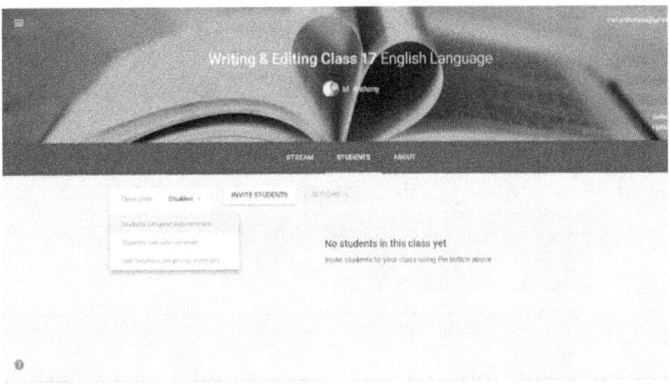

Using the 'Student's tab you can manage each student or several of them by managing their permission levels to access the classroom either by

- Giving the ability to post or comment
- Only comment
- Only teachers can post or comment

12-You can also send emails individually to students or even mute individual students from commenting.

Grading an Assignment

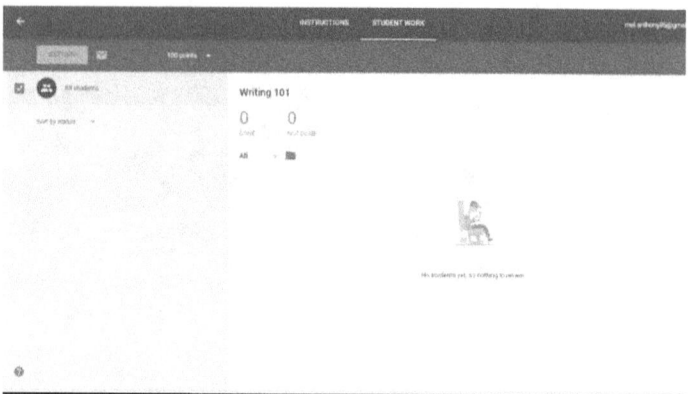

When you create an assignment and allow students to complete the assignment, it is time for grading. Click on the assignment you want to grade and you will see which student has completed it and who hasn't.

Classroom categorizes assignment as 'done' 'not done' 'late' or 'done late'.
The student's names will show and their files can be viewed to reveal the text field.

An assignment that has been received can then be graded and you need to click "No Grade" to grade a specific assignment. There is also a 'Points' section at the top of the page bar which you can change to give out a point system grading.
You can also download a student's work to your connected Google Drive folder to store.

Once a student submits a completed assignment to the Classroom, they are unable to make changes to that file unless you return it to them after grading.

Conclusion

And that is all there is to set up your Google Classroom! Easy to understand and quick to start up is the key to why Google Classroom is popular all throughout the world.

This is the basic features that a teacher needs to learn and understand in order to fully utilize Google Classroom. Getting familiar with the basics will help a teacher navigate the app and manage their classroom and students more efficiently. Once you have covered the basics, you can now start adding in different kinds of training modules or assignments and also creating differentiated assignments.

Chapter 2- Differentiating Assignments in Google Classroom

Google Classroom allows a teacher to create differentiation digital assignments based on a student's capabilities and levels of learning. Differentiated assignments are created based on a student's learning levels and skills with coping in the conventional classroom.

How to Give Assignments to Individual Students or Groups

This new feature is a very welcomed feature for teachers as it enables teachers to give out an assignment to exactly who they want whether it is for ONE student, a few students or even a group of students within the Google Classroom environment.

This feature gives teachers the flexibility in creating assignments for students based on the learning styles, their ability, reading levels and so on. So if a student is having difficulty reading or coping with general homework and assignments, the teacher would be able to conduct specialized assignments catered for this student to test out their learning capabilities. It also allows the teacher to give a student or a group of students extra assignments or assignments for a particular need.

To give different assignments to groups or individuals, all you need to do is use the drop-down menu to select the students that you want to send the assignment to.

Go to + Create Assignment > All Students > Drop-down list.
From here on, you can select the particular students or the groups you want to assign work to. Assignments can be anything such as independent study, assignments for extra credit as well as genius hour assignments.

Tips for Managing Differentiated Assignments in Google Classroom

1. Number your assignments

Numbering your assignments and giving them specific names related to the student or group you want to send the assignment to is imperative so that you don't get confused. Include a number as well as an identifier for your own sake and sanity. The great thing about Google Classroom's is that you can create as many assignments as you want and all of these assignments will be shown in tiles format via a clean and intuitive design interface.

2. Make your directions specific

The more information you provide when creating these assignments, the fewer excuses you'd receive from students for not completing their tasks. Also, make your instructions easy to understand. Have sufficient information given but make it concise and easy to comprehend.

3. Utilize a rubric

A rubric will be able to make your students understand the end goal of the assignment and your expectations of the quality of the task. Include a rubric if you have one of those and make it clear what your expectations and outcomes are for the task.

4. Designate a group leader

Yup! Just like the conventional classroom setting, team leaders come in handy for group activities via cloud-based learning as well. The team leader is tasked with creating new files and turning them in through the Classroom for the entire group.

Tips on Creating Differentiated Assignments on Google Classroom

Teachers using Google Classroom can transfer certain systems and methods that they use in the conventional classroom setting. When it comes to creating differentiated assignments, here are some tips on maximizing Google Classroom to reach your goal:

1. Focusing on the learning outcomes

Instead of an emphasis on the directions for each assignment, focus on the results of each assignment. You want to push forth the learning concept that you want students to demonstrate. You can offer them several choices of how they can get this done, sort of like choosing your adventure to complete a quest.

Make full use of the ability for commenting since students can privately speak to the teacher on an assignment, encourage them to write a comment to you on what approaches can they make and things that they can work on their assignments without revealing the content to other students.

2. Understand your learners

As a teacher, you already have a good idea of the various needs the students in your class has. Some students are ready for a challenge whereas some need some handholding. When creating your differentiated assignments, a good thing to do is to look through your class roster and tick out which students fit with which assignment. Determine if these assignments can connect to their needs and appeal or excite them. Because Google Classroom enables you to add in links from all over the web, take full use of it and add in a movie clip, a YouTube video or a link to a website explanation. This gives the students a better idea of what kind of outcome are you expecting for each task.

3. Maintain motivation with different challenges

If you give a student too hard a task, they will most likely give up rather than preserve. Differentiating assignments allows students to understand a topic and complete assignments that they can grasp, and this will eventually help maintain the motivation within themselves till they finally reach the level on par with the rest of the class.

4. Leveling

If you are worried that your students may want to opt for tasks and assignments that are below their learning ability level, then you can also choose to code the assignments with levels. Coding assignment options can help a student choose the right task just like how accelerated reader assists the students in selecting books that are at their level of ability.

Student Submission of Assignments

The beautiful thing about Google Classrooms is the interaction that a student has with it as well. Students can turn in their completed assignment in a variety of ways whether through a link, uploading a file to the Google Classroom or retrieving it from Google drive. When you request your students to upload their assignments, make sure to give a particular format for task uploading which includes the assignment number and their name.

Depending on the tasks, students can upload or send you their versions of assignment answers and solutions whether it is a blog post or a recreation of a city in Minecraft, a YouTube link and an essay even. All these things promote creativity in a student as opposed to the conventional paper format of writing and submitting answers and turning in assignments.

Teachers can accept a variety of different outcomes to assignments because Google Classroom collates all this information in one place for each student submission for a particular assignment. For example, for assignment A, students are required to link/post/upload or save an assignment for the Assignment A folder. If a student were to email a teacher to submit their work, their email could get easily lost, and the teacher would also have a hard time looking for it and also categorizing the assignment.

Privacy

Student's submission is not publicly available for all other students to see. In a conventional classroom setting, it is very easy to know which student submitted what especially if each submission is a different thing altogether. Via Google Classroom, students with alternative answers and options will be able to send in their assignment in privacy and not have to worry about being called out in front of the whole class.

Conclusion

Google Classroom provides a new and different interaction between student and teacher by way of back and forth private feedbacks and commenting as well as discussion on a document.

It may sound like the teacher is then available 24-7 to answer questions that the student may have but scheduling, as well as communicating hours, can be established so that students can connect to the teachers at these particular times.

Turning in assignments through Google Classroom also gives the students the major advantage of finding solutions or completing the assignments in various ways. This means it cuts the traditional and conventional method of learning and allows the student to dig deep into their creativity chamber to find the means to solving a task according to their passions, interest, and needs. Differentiating tasks also allows the teacher to ensure that students who are falling behind will be able to catch up with their classmates with extra coaching, feedback and assisting the student whenever possible.

Chapter 3- Best Practices Of Google Classroom Teachers Should Know

Google Classroom allows you to extend the blended learning experience in a variety of ways and come 2017; teachers can create excellent number of ways to enhance a student's grasp of school subjects and increase learning capabilities. The possibilities are endless where Google is concerned.

Google's biggest asset is its simplicity and ease of use. Using the various Google applications doesn't require a textbook to learn it, as with Google Classroom, all other apps are simple to set up, quick to learn and saves time and energy to get things done and organize your various files and documents. In this chapter, we will share ten best practices for Google Classroom that you can employ, to fully make use and take advantage of this pioneering online education tool.

1. Reduce the carbon footprints of your class

The idea of Google Classroom is to make things easier for teachers and students alike when learning things. It takes the conventional classroom and places it on the online sphere and enables students and educators to create spreadsheets and presentations, online documents and it makes sharing and communicating easier. Creating and sharing things digitally eliminates the need to printing. Schools use a lot of papers but utilizing Classroom enables you to remove the necessity of paper for simple things. Have an assignment? Save some trees, time and money by creating them on Classroom, distributing it to your students in your Classroom.

2. Distribute and Collect Student's homework easily

The whole point of creating the assignments via Google Classroom is so that you can distribute it and collect the assignments quickly. Yes, you can say that you could get it done via email too. But Classroom's enable all these things to be done in one place. You'll know who has sent an assignment, who have passed their deadline and who needs more help with their work. It's all about lessening the hassle in your life.

3. Utilizing the feedback function

With instant access, teachers are able to clarify doubts, concerns and misconceptions their students may have by providing feedback as and when students need it. As teachers, you eliminate possible issues that might arise while students are doing their assignments. This reduces the headache you might have upon receiving the assignments that don't meet the requirements. Assignments that are handed in that have issues can be immediately rectified as well, through private one-on-one feedback with the relevant student.

4. Create your personalized learning environment

The main benefit of Google Classroom is the freedom that it gives teachers. Very often, teachers are required to follow the national syllabus forwarded by the Department or Ministry of Education in a country. While this is rightly done for the sake of uniformity and to ensure students across the country have access to the same level of education, utilizing Classroom, on the other hand, gives teachers the freedom to add and create a different environment for learning.

Teachers can focus on using different materials, subjects, and cater to the different levels and needs of students. If you are using Google Classroom, then make sure you use this aspect to your fullest advantage. You would be able to endorse a personalized learning system by giving your students different learning preferences such as choices of submitting answers, various types of online assignments and using online resources.

5. Encourage real world applications

Encourage students to submit their assignments using real world material whether it's a series of videos or photos, a compilation of multimedia applications, using the many different apps out there to create amazing online presentations are just some of the things that students can do that will increase their learning tendencies and spark online discussions within the Classroom. This enables the students to apply and implement assignments that they have done in their real lives.

6. Allow shy students to participate

As teachers, we know which students are more extrovert that the other. Sometimes in conventional classroom settings, the shy kid or the kid with self-esteem issues or those that lack confidence have problems participating in classroom activities, speaking out or even raising their hand to answer questions. Google Classroom gives a safety barrier for students that fall into this category but allowing them to be more open with discussing and expressing themselves. As the teacher, you can also find creative ways to encourage these students to open up via game-based learning to promote trust, openness, teamwork, and collaboration.

7. Allow for coaching

Some students need more coaching and a little bit more explanation. If you know some students in your class that needs it, you can give them extra instructions by privately messaging them. You can always follow up with them while they are doing their assignments just to check if they are on the right track. Additionally, you can also invite another teacher to collaborate and help with coaching your students.

Interactive Activities Using Google Classroom

The more and more you use Google Classrooms, the more you will be able to use Classrooms in many more ways than just connecting with your students and creating assignments.

Google Classroom, combined with other Google products such as Google Slides can really deliver powerful interactive user experiences and deliver engaging and valuable content.

Teachers looking to create engaging experiences in Google Classroom can use Google Slides and other tools in the Google suite of products to create unique experiences.

Here are some exciting ways that you can use Google Classroom and Google Slides to create an engaging learning experience for your students:

1. Create ebooks via PDF

PDF files are so versatile and you can open them in any kind of device. Want to distribute information only for read-only purposes? Create a PDF! You can use Google Docs or even Google Slides for this purpose and then save it as a PDF document before sending it out to your classroom.

2. Create an slide deck book

Make your textbooks paperless too, not just assignments. Teachers can derive engaging and interactive content from the web and include it in the slide deck books, upload it to the Google Classroom and allow your students to access them. Make sure to keep it as read only.

3. Play Jeopardy

This method has been used in plenty of Google Classroom and the idea was created by <u>Eric Curts, a Google Certified Innovator, created this template that you can copy into your own Google Drive to customize with your own questions and answers.</u> Scores can be kept on another slide that only you can control.

4. Create Game-Show Style Review Games

Another creative teacher came up with a Google Slide of '<u>Who Wants to be a Millionaire?</u>'. The template allows you to add in your questions and get students to enter the answers in the text box. Again, you keep the score!

5. Use Animation

Did you know you can create animations in your Google Slide and share in on Classroom? This <u>tutorial</u> shows you how. You can also encourage your students to create animation to explain their assignments. This is really making them push boundaries and think out of the box.

6. Create stories sand adventures

Using Google Slides and uploading them to Google Classroom to tell a story. Turn a question into a story and teach your students to create an adventure to describe their decision for the outcome of the character in their story. The stories can be a certain path that the students have chosen for the character or a story that explains the process of finding a solution.

7. Using Flash Cards

Flash cards are great ways to increase the ability to understand a subject or topic. Do you want to create an interactive sessions on Google Classroom using flash cards? You can start by utilizing Google Sheets which gives you a graphic display of words and questions and then to reveal the answers, all you need to do is click. Compared to paper flash cards, these digital flash cards allow you to easily change the questions, colors as well as the answers of the cards depending on what you are teaching the class. Digital flash cards also are an interactive presentation method that is guaranteed to engage your Classroom and bring about a new way of teaching using Google Classroom's digital space.

Make vocabulary lessons, geography lessons and even history lessons fund and entertaining with digital flash cards. Here are two great resources that you can use to help you create your very own set of flash cards:
https://www.youtube.com/watch?v=tPuUc--xHto andhttp://sites.godfrey-lee.org/google-docs/spreadsheets/flashcards

8. Host an online viewing party

Get your students to connect to Classroom at a pre-determined date and time when there is a noteworthy performance, play or even movie that is related to the subjects you are teaching in your class. Let them view the video together and also interact with them by adding questions to your Google Classroom and allowing your students to reply to you in real time. This way, you can see assess them on their reflections, level of understanding and their observations. You can also give your own interpretation of the scene and explain it again to students who do not quite understand.

Conclusion

There is no limit to what a teacher can do with Google Classroom and the entire Google suite of apps whether its Google Slides or Google Calendar or even Google Maps. The only thing you would need is creativity and the desire to give your student a different experience when using Google Classroom.

Chapter 4- Great Apps to Use Together with Google Classroom

As of 2017, Google Classrooms can be accessed without the need for a unique G Suite for Education ID. That said, having an ID for the suite and accessing the site via the ID helps keep things organized in the online sphere. It ensures that you do not mix your private and personal documents and information into your Google Drive or Gmail account connected to your suite.

For the full list, you can check out <u>Google for Education Products section</u> and add on any other apps which you feel will help address the needs and requirements of your Google Classroom.

But for now, this chapter will focus on the most resourceful and convenient apps that do plenty of things with Google Classroom.
Here they are!

1. American Museum of Natural History

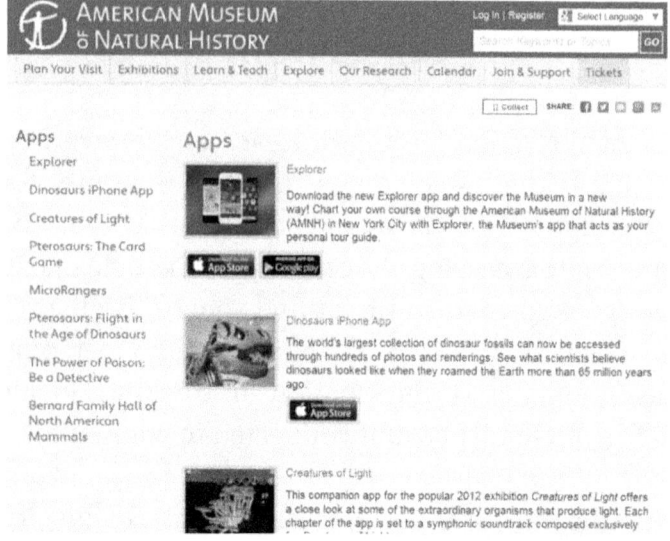

Click into this website and you'll be spoiled for choice. There are plenty of apps that you can download and integrate with Google Classrooms. From adding on apps on Dinosaurs to the Solar system, doing detective work, the AMNH has got you covered. Install any of these apps to gain access to archival photos, curator commentary and many more.

2. cK-12

You can download this app either in student mode or teacher mode. If you plan on creating a differentiated assignment, then this website is your Holy Grail because it is filled with a library of online textbooks, flashcards, exercises videos and all of it is for free!

3. Alma

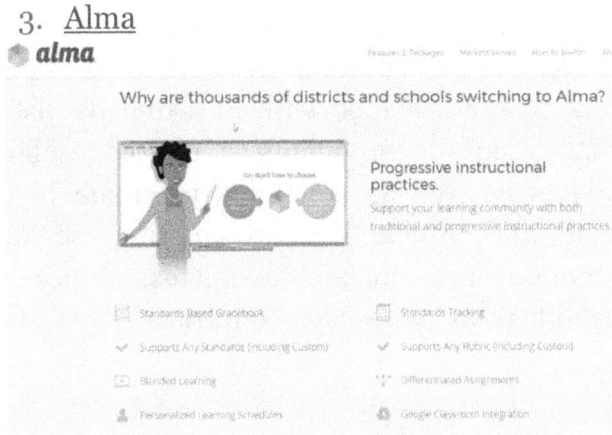

Alma is a cool and sleek software designed to help schools and teachers better their school management and learning management system and student information system. Its interface is user-friendly and it has systems with grading, standards tracking and supports any kind of Rubric.

4. Buncee

Encourage your student's creativity through Buncee, a presentation tool that is highly interactive and loaded with an extensive list of visualization components. Buncee allows students as well as educators to create highly visual and interactive presentation stickers, animation and built-in templates. Buncee is currently used in over 127 countries.

5. Google Cultural Institute

The Google Cultural Institute features an online collection of art, exhibits, and archives sourced from around the world. Need to link an assignment with a content? Look it up on Google Cultural Institute. You can find an extensive list of topics and articles categorized under experiments, historical figures and events, movements as well as artists curated from museums and archives worldwide.

6. Curiosity.com

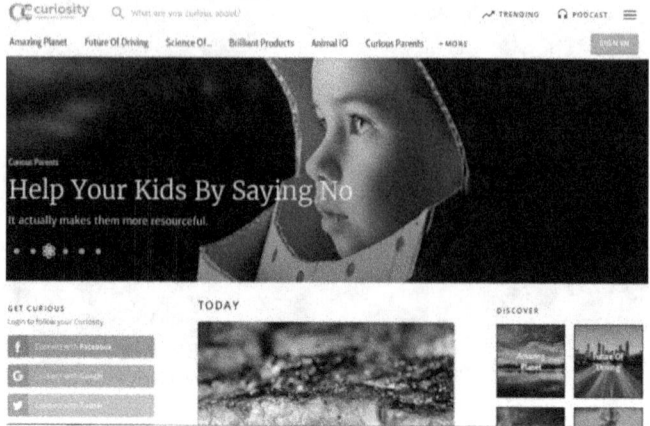

With the goal to ignite curiosity and inspire, this app curates and creates content for millions of learners all around the world. Editors look for content and present it in the best way possible. Curiosity can be accessed through the website or on their app.

7. Discovery Education

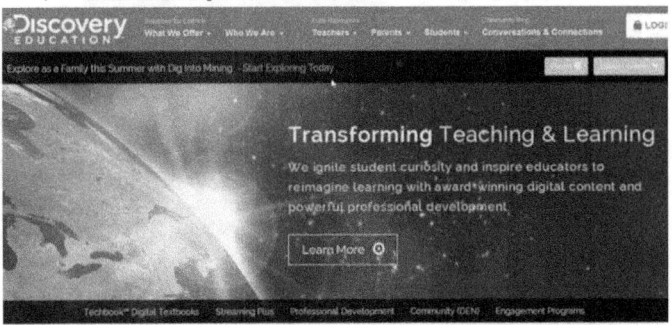

Techbook™ Digital Textbooks

Discovery Education is another source of well curated information loaded into digital textbooks, digital media and Virtual Field Trips that feature content that is relevant and dynamic. They also have easy-to-use tools and resources that enable teachers to include it in their differentiated learning modules to improve their student's achievements.

8. DuoLingo

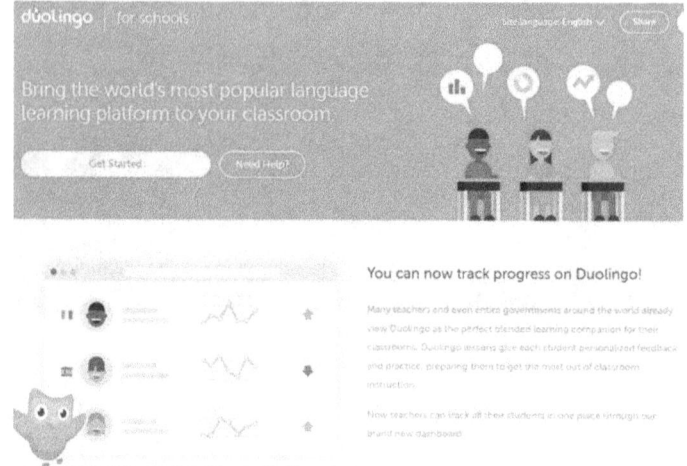

By far the world's most popular language website, DuoLingo for schools is the ideal blended learning companion for their classrooms all around the world. Duolingo lessons give personalized feedback and practice to each student, preparing them to get the most out of classroom instruction.

9. EdPuzzle

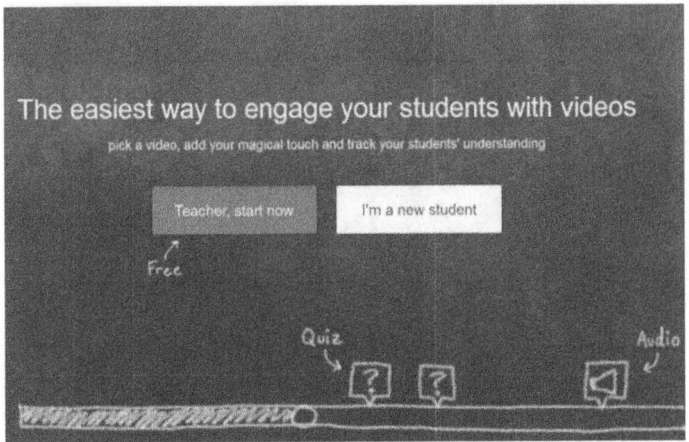

To be used by both educators as well as learners, EedPuzzle allows you to create your own videos and include interactive lessons, voice over, audio and many more to turn any video into a lesson. What's more, teachers can also track if a student watches the videos, the answers they give and how many times they view a video.

10. Edulastic

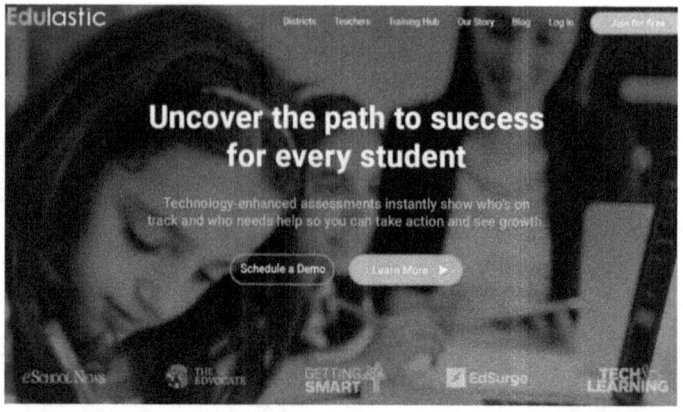

Edulastic is a platform that allows for personalized formative assessment for K-12 students, teachers and school districts. It gives educators a highly interactive, cloud-based learning environment and gives deeper insights into students' understanding of a subject.

11. Gale Cengage Learning

Looking for library content? Look no further than Gale, a Cengage Company that has extensive partnerships with libraries around the world, enabling both students and teachers to gain access to datases, ebooks, manuscripts, Associated Press content and online archives.

12. Geogebra

GeoGebra is an excellent app for both educators and students alike. It includes a graphing calculator, 3D calendar and geometry calculator that can be used to produced geometry, calculus, statistics, 3D math and functions.

13. Learn Zillion

This website is the world's first curriculum-as-a-service program that utilizes digital curricular materials and combine it with an enterprise platform as well as professional services to enable districts and states to effectively manager their curricula and provide their teachers with the best tools to make engaged and blended learning possible

14. Listenwise

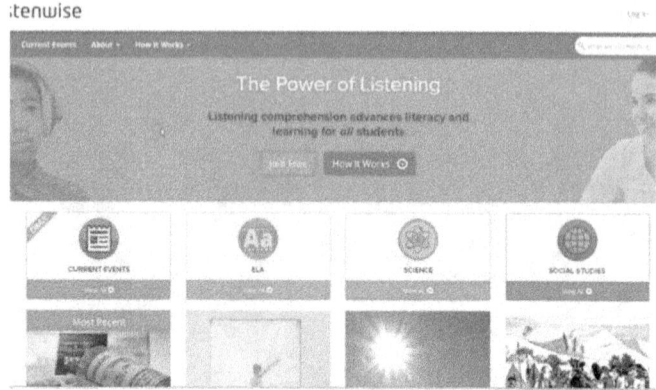

This listening skills platform harnesses the power of listening to empower learning and literacy for students. This site features podcasts and public radio content. Teachers and teach and assess listening skills using this software. Listenwise is an award winning program.

15. LucidPress

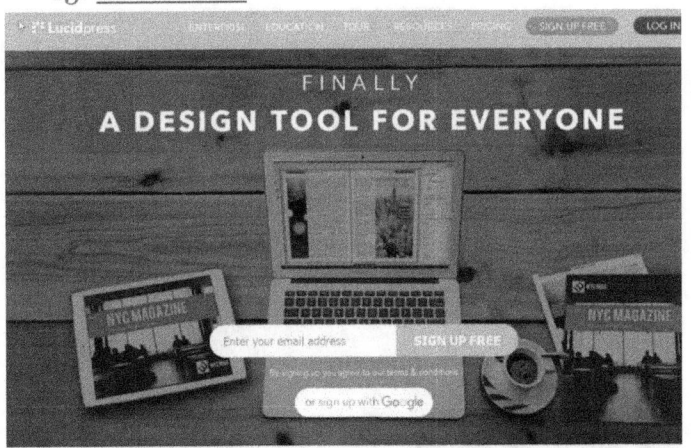

**QUICKLY CREATE AND SHARE
STUNNING VISUAL CONTENT**

Encourage your students to create visually stunning materials for their assignments using Lucid press. From newsletters to brochures, digital magazines to online flyers, LucidPress incorporates an intuitive interface of easy drag-and-drop that is easy for beginners and also for experienced designers.

16. Nearpod

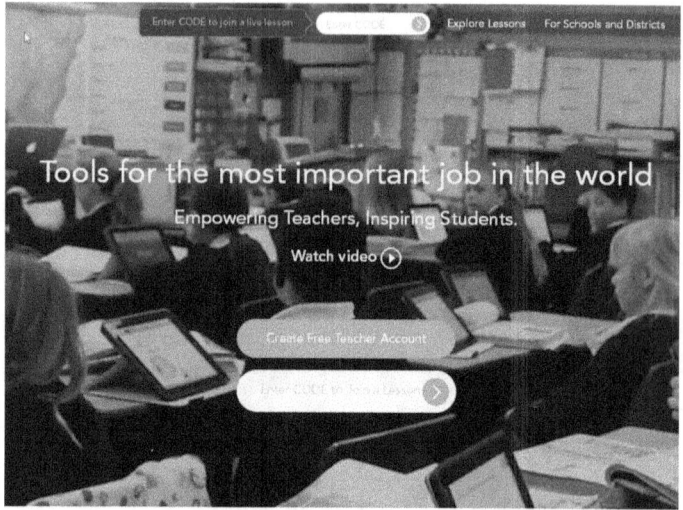

Create intuitive lessons with Nearpod whether in ppt. or jpeg. or PDF files and upload them to your Google Classroom. Nearpod enables teachers to create mobile presentations and share and control the presentation in real time.

17. [Newsela](#)

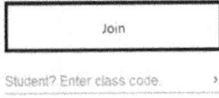

With Newsela, you can integrate articles into your assignments with embedded assessments. Start a dialogue, customize prompts and facilitate close reading with this app.

18. PBS Learning Media

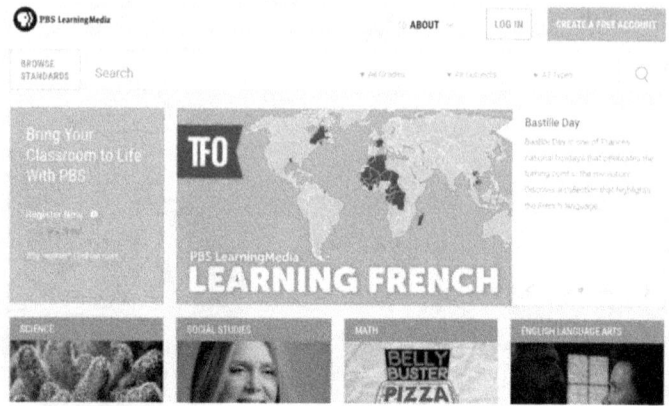

The PBS Learning Media is a standards-aligned digital resource that gives educators and students access to digital resources both for student and professional development. With PBS Learning, educators can fully utilize a student's digital learning experiences.

19. Pear Deck

This tool enables each student in your class to connect to your presentation on any device, and when connected, they can answer your interactive questions, and learn from their peers simultaneously. If you have a previous presentation that you want to enhance visually, the export it to Pear Deck and it will be transformed into powerful and visually appealing presentations.

20. Quizizz

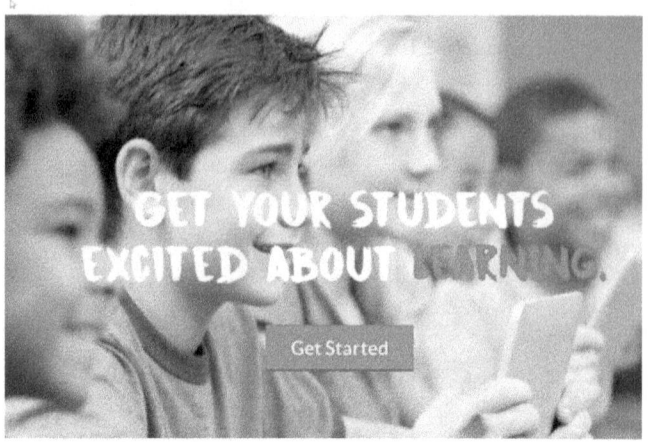

With built in avatars, music, memes, leaderboards and themes, Quizizz enables a teacher to easily create engaging quizzes and it can be uploaded to Google Classroom. Teachers can also obtain student-level data while the quizzes are being played.

21. RMBooks

If you conduct reading and writing classes, then this add-on is for you. RM Books is a ebook solution designed specifically for schools. It is a free-to-use service that requires no upfront payment. Students can have access to digital textbooks, classic literature and new releases from a wide array of genres.

22. Science Buddies

Get connected to thousands of resources for your student's next science project from convenient kits to summer science camps, science blogs and many more. Explore how to make underwater fireworks, making a paper fish swim and even learn how to build a balloon-powered car.

23. Texthelp

Another excellent tool for reading and writing classes. Texthelp can be used in Google Classroom as a support tool for languages, reading, comprehension and writing. It also has a dyslexia module that would be useful for those with reading and writing difficulties.

24. Versal

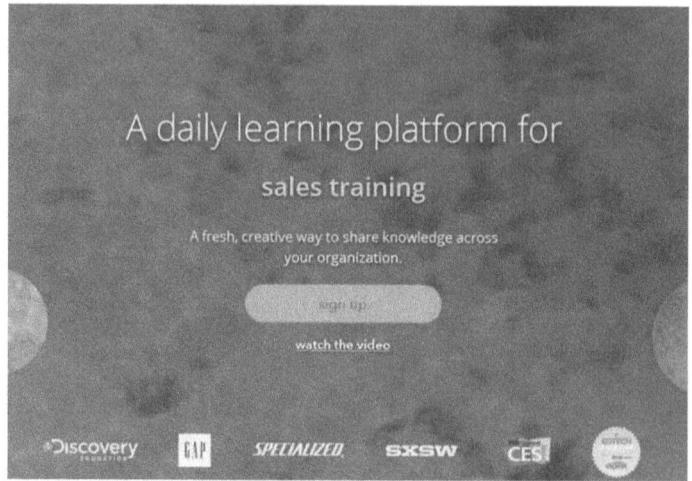

If you are using Google Classroom for professional development, then this website is for you. Versasl is a platform everyday learning, with its content geared to helping companies create a vibrant culture of collaborative knowledge sharing.

25. WeVideo

WeVideo is the online video editor that makes it easy to capture, create, view and share your movies at up to 4K resolution for stunning playback anywhere.

Can be used for higher education, life, school, students as well as teachers, WeVideo takes the video sharing to a different level by empowering people to share their stories, with powerful video editing features and create story-telling video formats.

26. IXL

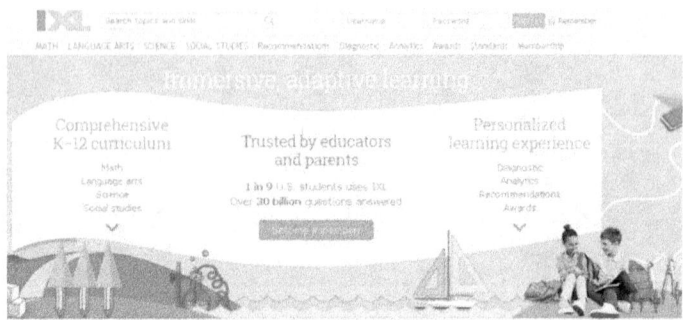

IXL Learning is a company dedicated to inventing new educational technologies that help in the teaching and learning process. Educators can use this website to gain plenty of algorithmically generated questions, have access to real-time analytical reports and scoring whenever students are given quizzes, assignments and tasks.

27. Little Sis

management model for Google Classroom.
- **The Classroom Explorer** gives global insights into Classroom adoption and allows he bulk administrative actions on classes.
- **Sync Jobs** auto-create classes in Google Classroom and update class rosters from student information exports, making it easier for schools to adopt and maintain Google Classroom at scale.

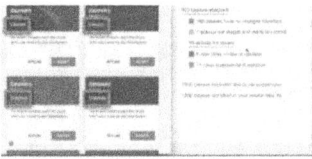

Analyze and sync guardian invitations

For districts that have effective processes to centralize the collection of parent email addresses, syncing Classroom guardians can be a powerful tool for family engagement and school-to-home communication. *Whether or not you choose to sync your classes and rosters*, Little SIS can:
- Analyze the current state of all guardian invitations on the domain.
- Manually or automatically send guardian invitations for eligible students from an SIS export.

This program makes things easier for teachers by auto-creating classes in Google Classroom and auto-syncs class rosters directly from student information exports. This makes it easier for schools to adopt and maintain Google Classroom at a bigger scale. It also helps teachers that manage two or more classes.

28. Go Guardian

The essential tool for Chromebook schools.

In the classroom or off-campus, GoGuardian makes teaching and technology management easier, safer, and more fun.

Fast

Start seeing real data from across your entire district immediately.

Easy

Users with Google Admin Console access can download and deploy the extension in minutes.

Secure

We use AWS and TLS-encryption to protect your data in storage and transit.

Trusted

Used by thousands of districts to protect millions of students worldwide.

GoGuardian can be accessed by Administrators of a School or Teachers. For teachers, GoGuardian enables educators to align the classroom that they have set up and conveniently sync student's information such as enrollment, subject and class period across platforms.

Conclusion

These are just some of the amazing apps and websites that can be synced or integrated with Google Classroom. As more and more schools across the world use Google Classroom, the need for more apps will rise. Teachers are encouraged to continuously look out for interesting apps that can help them make their work easier, lighter and more organized.

Chapter 5- Creative Ways to Use Google Classrooms to Teach

Teaching Math

If you are thinking how else you can expand the experience of learning math or using Classroom in your math classes, here are some creative ways to build on.

1- Problem Of the Week

Aptly known as POW, POWs can be anything that you feel needs more attention. It can be a problem you have identified or a problem that your students can identify. You can create games that can help students learn about the problem differently and participating students can submit their work directly to Google Classroom.

NUMBER LINES

1. **Make each line add up to 16.**

   (2)—(5)—(3)—( )
   ( )   (2)
   ( )—( )—(2)—( )

2. **Make each line add up to 20.**

   (9)—( )—( )—(4)
   ( )   ( )
   ( )—(3)—(14)—( )

2- Link Interactive Simulations

There are several websites dedicated to providing helpful math simulations. Sites like <u>Explore Learning</u> have thousands of math simulations and math variations that students can look up to solve mathematical problems. You can link these URLs in your classroom either as part of an assignment or through an Announcement.

3- Link to Playsheets

Playsheets fall between gamification and GBL. Teachers can link up relevant <u>Playsheets</u> and give these are assignment for students. These playsheets give immediate feedback to students and it is an excellent learning and motivational tool that tells the students that they are on the right track.

4- Use Google Draw

<u>Google Draw</u> is another creative tool that allow students and teachers to create virtual manipulations such as charts, Algebra tiles and so on. Draw images that make it easy for students to identify with Math. This can be used to create differentiated assignments targeting students with different learning levels.

5- Use digital tools

Digital tools such as <u>Desmos</u>, <u>Geogebra</u>, <u>Daum Equation Editor</u> can also be used to solve various math problems. These tools can be used from Google Drive and integrated with other Google documents. Once done, students can submit their solved problems to Google Classroom.

GOOGLE CLASSROOM

GeoGebra Math Apps

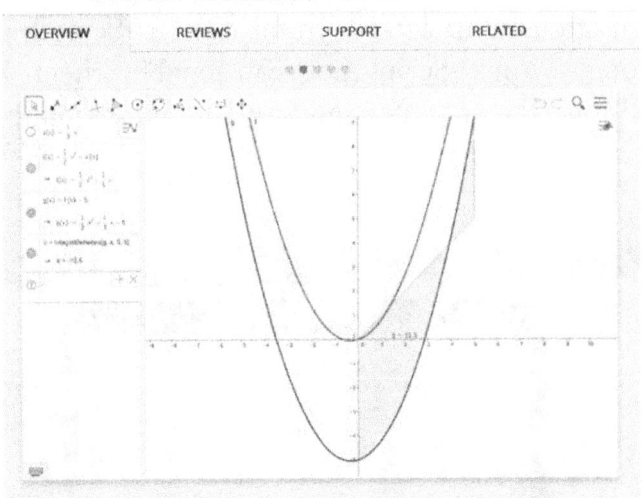

Teach programming

Get students to use programs such as <u>Scratch</u> or <u>Google Apps Script</u> that can enable them to exhibit their understanding of mathematical concepts.

Teaching Science

1- Hangout with Experts

Get an expert you are connected to in real life to talk about their experiences either working in a science related field or to help students with science related subjects. You can use Google Hangouts to send questions the class has and link it to your Google Classroom. This enables the students to access the Hangout and participate in the questioning or even watch the interview after the session is done. The Hangout Session can be archived for later viewing.

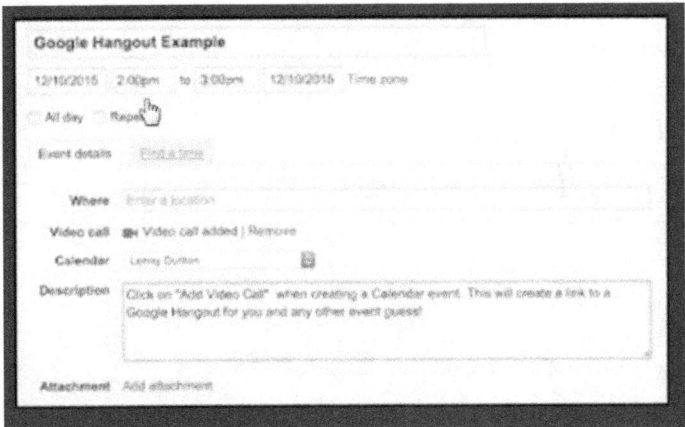

2- Collecting Evidence

Have your students submit 'evidence' of science experiments by sending it photos or videos of their science projects and upload it to Google Classroom.

3- Give Real life examples

Tailor-make your science projects and assignments so that it gets students going out to get real life samples which they can record on their mobile devices. They can take these images and submit it immediately to the Google Classroom. Make it interesting- students that submit their answers faster get extra points!

4- Crowdsourcing information

Get students into the whole idea and activity of crowdsourcing. Create a Google Spreadsheet with a specific topic and specify what information they need and what goals the project needs to accomplish. Upload the document to Google Classroom and get students to find and contribute information.

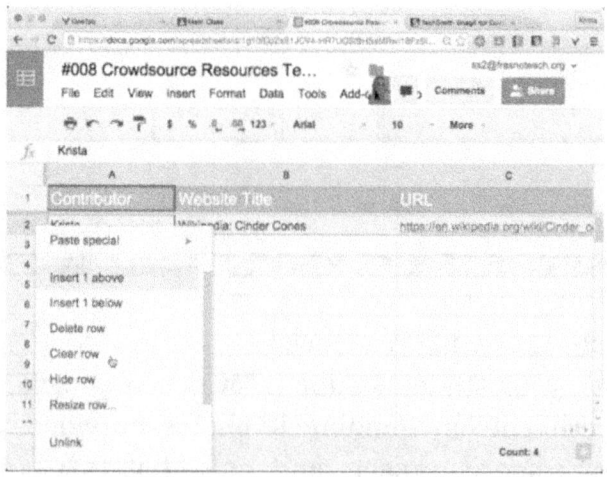

Teaching Writing & Reading

1- Provide Templates

Allow students to access writing templates on Google Classroom for things such as formal letters, informal letters, report writing, assignment templates, resumes and cover letter formats.

2- Reading Records

Establish a <u>weekly reading record</u> on Google Classroom where they can record information on the times that they have read during the week. So instead of writing it down on a reading diary, allow them to update a form on Google Classroom by entering the necessary data. This allows them to immediately add in the information of the books that they have read while it is still fresh in their minds.

Class Reading Record

Tom Barrett
ICT in my Classroom - http://tbarrett.edublogs.org

Name

Date

Book Title

Page Numbers

Comments
What did you enjoy? Did you struggle with any words? What help did you get?

I read with...

Submit

3- Collaborate on Writing Projects

Get your students to collaborate on writing projects via Group assignments. These projects can be anything from preparing newspaper articles, journals, e-portfolios and so on.

4- Spelling Tests

You can create a simple 1-10 or 1-20 <u>weekly spelling test</u> via Google Form. Get students to type in their answers as you read out the list of words. Once completed, apply formula to judge if they are correct or not and it becomes self marking.

Spelling Test

Tom Barrett
ICT in my Classroom - http://tbarrett.edublogs.org

Name

1)

2)

3)

4)

5)

6)

8)

Teaching Physical Education

Didn't think PE could be done via Google Classroom? Here are some ideas:

1- Post Fitness Videos

Post fitness videos to help your students understand how to perform a workout. Send out videos to any psychical activity that you want students to conduct on their next PE session or you can also just post a video after classes so students can practice the exercise in their own time and work on their form.

2- Get students to post videos of their daily workout

Have your students post videos in the public feed on your Google Classroom with a hashtag such as #midweekfitspo. Encourage students to workout and post their videos each week.

3- Link to safety videos

Post up safety videos for your PE activities so your students know what kind of skills that they need to follow in order to exercise safely.

4- Post Resources for activities

PE teachers can also post up useful resources for games and activities such as rules and method of playing ahead in time before the student's next PE session. It would help the students prepare and know what to expect for their next class.

5- Create a Fitness Tracker

Assign students to a **Fitness Tracker spreadsheet** and make a copy for each student. Assign a due date for the end of the semester for their physical education class. You can monitor each student's progress by checking out the assignment folder in the Google Classroom.

Use the spreadsheet to get your students to track their progress. Whenever students update their results, the spreadsheet automatically updates to dynamic charts so students can see their progress visually over the entire semester.

You can either pair students up to work in partners or individually. Get the students take photos of each other's forms when practicing certain tasks so that you can evaluate their form and correct it by way of giving them feedback via Classroom or during PE classes. A rubric would be helpful here too so that students can self-evaluate their own workouts and make corrections where necessary.

Other Teaching Methods to Use

1- Attach Patterns and Structures

Upload patterns and structures that students and identify and explain. Students can also collaborate with other student to identify patterns and structures to come up with solutions.

2- Use geometric concepts

Use Google Drawings or Slides to insert drawings of geometric figures for maths, science and even for art.

3- Collaborate online with other teachers

If you know other teachers have modules or projects which would come in handy with your class, collaborate together and enable your students to join is as well. Different teachers allow for different resources and the teaching load can also be distributed.

4- Peer Tutoring

Senior students can also be allowed to access your Google Classroom at an agreed time on a weekly basis to tutor and give support to junior students or students in differentiated assignments.

5- Celebrate success

Google Classroom also enables the teacher to encourage students through comments whenever they submit and assignment because feedback can be given immediately and this can be done either private or publicly

6- Digital quizzes

Quizzes can be used for various subjects on Google Classroom. Get your students to submit their answers quickly for extra points.

7- Share presentations

Share whatever presentations and slides that you have with your students to help them with whatever assignments that you have given them.

Conclusion

These are some of the things that you can do with Google Classroom that are subject specific. As Google continues to update and enhance their products, there will be even more ways to use digital tools to heighten the experience of learning

Chapter 6- 10 Things Students Can Do Using Google Classroom

Google Classroom was built for both the educator and learner in mind. It isn't only the teachers who can do so many things with Google Classroom, but students can also harness the full capabilities of this application. Student's reaction to Google Classroom is whenever the teacher, who is the main Manager of the Classroom, uploads content in the Classroom.

Here are some of the various things that students can do with Google Classroom.

1. Change Ownership

When you turn in an assignment, the teacher becomes the owner of your document. You are no longer the owner, and therefore you are unable to edit the text. Turned in the wrong assignment? Simply click on the 'Unsubmit' button. You would need to refresh Google Classroom once you un-submit so that you can resend a new document.

2. Assignment listings

Students can find a list of all the assignments created by teachers by clicking on the menu icon located at the top left-hand corner of Google Classroom. Practically all assignments that have not been archived can be viewed in this list.

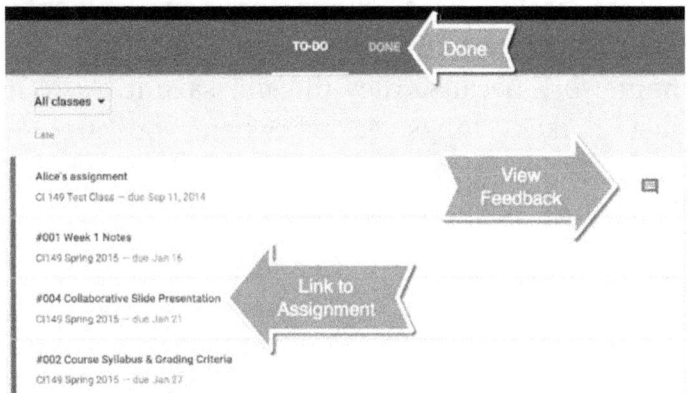

3. Utilize the Mobile App for easier access

We know students are always on their mobile phone. One of the best ways to get notified if you have a new assignment is through the Google Classroom's mobile app. The mobile app can be downloaded and installed from the Playstore or iTunes. The app allows students to view their assignments and submit their work directly from the app. This mainly works when students are requested to submit real life samples, or a video or a combination of photos. All they need to do is take pictures of their samples or their solutions and then upload it to the Google Classroom.

4. No worries if you haven't clicked on Save

Encourage your students to use Google Docs to do their assignments. If you have given work that requires them to write reports, write a story or anything that requires their use of a Word document, use Google Docs because it saves edits automatically. This eliminates your student's excuses of not being able to complete their homework because they did not save it. Also, it just makes things easier when you are so engrossed with completing your work, you forget to save; Google Docs does it for you.

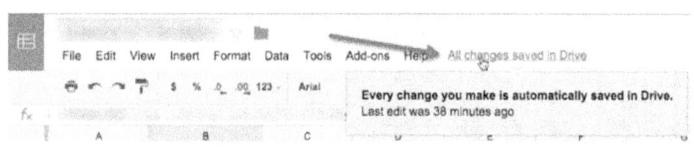

5. Sharing isn't the same thing as turning in

When a student clicks open an assignment to hand in their assignment, they need to click on TURN IN. Sharing an assignment to the Google Classroom is not the same thing as turning in your completed work. Make sure you click on TURN IN to submit your assignment in due time.

6. You will not lose assignments

Unless you delete it. Any documents you upload to your Google Classroom is only seen between you and the teacher. Any assignments you upload to your Google Drive will be seen on the teachers Google Drive as well. Your Google Drive is the storage system for Google Classroom and it works the same way for both the teacher as well as the students.

7. Due Dates

You'd have a harder time explaining to your teacher why you have not submitted your assignment especially since the due dates are continuously shown on an assignment. Assignments that are not due yet are indicated on the class tile on the home page as well as the left of the page late assignments have a particular folder, where the teacher can accurately see the assignments listing from the menu icon on the upper left of the page.

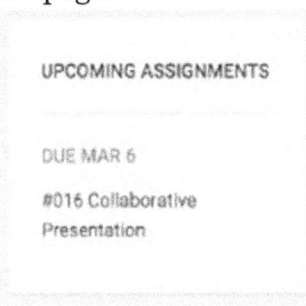

8. Returning an Assignment

Students working on a Google Document can return at any time to the file that they are working on. Get back to the assignment stream and click on Open and it will take you to a link to the documents that you have on Google Drive. Click on the document and get back right into it. You can also access this file directly from your personal Google Drive. It is the same way you click on any document on your desktop to work on it again. Plus side is Google Docs auto saves.

9. Communicating with teachers

It's either you communicate publicly on Google Classrooms for the entire class to see, or you communicate privately. Communicating privately helps a lot especially for students who are shy and prefer to speak to the teacher directly without the involvement of other classmates. It also helps the teacher speak privately to address a student's issue on an assignment without making them feel inadequate or that they have not done well.

10. Commenting on Assignments

Comments on an assignment are viewable by your classmates on Google Classroom when it is made on any assignments uploaded to the app. Students just need to click on 'Add Comments' under an assignment. If students would like to communicate in private, with you, they can leave it on the assignment submission page. Within a specific document, you can use the File Menu and click on 'Email collaborators' to message or link a document to the teacher.

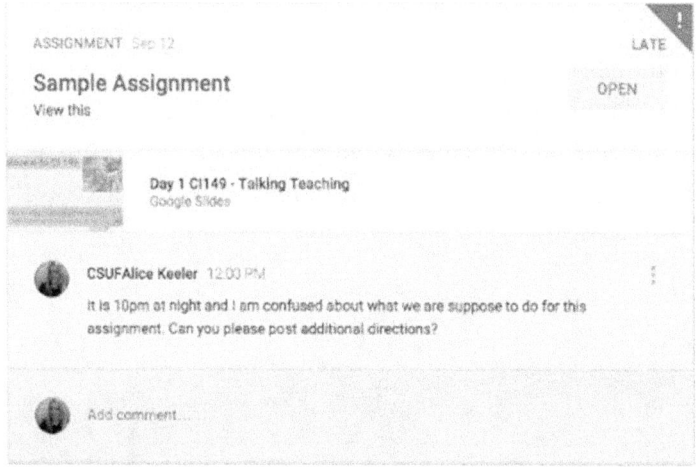

11. Add Additional files to an assignment

Students and teachers can both add additional files to an assignment. For students, they can add in files that did not come together with a template the teacher gave. You can click on ADD additional files on the assignment submission page again. Links from websites can also be added. Additional files help in the attempt to provide a wholesome blended learning approach in schools because you can add files of different formats and types.

Conclusion

Google Classroom has two different 'views' built. One for the educator and one for the learner. Knowing what students can do on their end can help a teacher navigate Google Classroom's easier. It helps to know what the student can see so that if they have any issues accessing the site or finding out how to view assignment, you will find it easy to help them navigate through Classroom.

Chapter 7- Using Google Classroom for Professional Development

The possibility for Google Classroom is endless, especially since anyone owning a Google account can start a class and teach to anyone they want on whatever topic they want.

Teachers, educators, lecturers and school administrators alike, the primary clientele of Google Classroom have also been using Google Classroom for Professional Development purposes too.

Here is an example of how you can apply Professional Development into your administration.

Professional Learning Communities & Google Classroom

1- Creating the Course

The idea of the Professional Learning Communities begins first by setting up a course. If your school has a head of the grade or Team Lead, then the Team Lead can create a class and then invite the rest of the team members to join the course. Those who are interested can be given a code or the Team Leader can send a link.

2- Number of courses

The team lead can open one course per subject or one course per grade. This is entirely up to the way the school system has been structured. If the school has departments based on Subjects, then the Team Lead can create courses based on these subjects. For example, the Team Leader can create courses for Geography and one for Minerals or one for mathematics and only one course for Science. The Language Team Lead can create a course for Writing, one for Reading and maybe one for Social Science.

3- Initiating Discussions

Discussions can be initiated via the Announcement Function or the Assignments function. Before meetings, the Team Lead can attach the agenda so everyone knows what will be discussed and members can also give input to discuss in real time. Announcements can be used for quick highlights or to spread a message about an event, such as 'Date for Inter-School Athletics Meet-up' whereas assignments can be used exclusively to give out tasks, delegate work and so on.

4- Take meetings online

Because of the comment feature, the Team Leader can also decide if meetings can be conducted via Google Classroom instead of meeting in person. Members can also use the Classroom Course to co-create common assignments in real time, even when members are at different locations.

Using Google Classroom for Training Sessions

A trainer or teacher can use Google Classroom for training courses as well.

1- Sending out course outline

The instructor can send out a course outline via email to the entire department and allow teachers who want to participate to enter the code to join the class. OR the trainer can also send the class code to a particular group of teachers whom the instructor feels would benefit from the course.

The course outline also needs to mention how long the training is for and how often do the teachers need to come online to discuss/collaborate/receive assignments. The course must also state when the teachers need to submit or answer or participate in the class in real time.

2- Handing out assignments

Based on the course outline, the trainer can publish assignments via the Assignments tab. From there on, use the option of "Make a copy for each student" to distribute the document to the entire class.

3- Collaboration

The trainer and learners can answer individually or work in groups to finish up the assignment given, based on the course outline. These collaborations can happen in real time if the course outline has stated when and what time should all learners come online. But of course, there will also be other assignments and opportunities where learners and the trainer can do this at their leisure.

Some training methods involve participation in real time to produce evidence of the knowledge that the class has gained during training. Your course outline should mention this.

Principal – Teacher Collaboration with Google Classroom

Collaborations between the Principal and administrators of a school can also take place in the Google Classroom sphere.

1- Guidance & Instruction

Principals can also forward instructions or plan events via Google Classroom with other coordinators of the school. Google Classroom creates a secure and private online avenue among principals and school teachers and administrators to share important details of information regarding school exams, assessment tests and so on.

This gateway can also be used to share information with regards to the DOE information. Principals can also create short assessments for teachers to complete for increment purposes or promotion reasons.

Google Classroom can also be used on for Professional Development Days where teachers can collaborate in real time with experts and professionals and perform vertical planning sessions, construct comprehensive assessments collaboratively, and support each other in various ways.

Professional Learning Possibilities

Google Class brings about the ability to gather a group of professionals to collaborate on issues or to work on solutions. Think of it as an online brainstorming session in a highly intuitive online avenue. So what other learning possibilities can be done with professionals using Google Classroom?

1- Book study
   Study a book or a journal together online. This works ideally for anyone wanting to discuss matters in details but do not want to meet at a physical location. Anyone from around the world can be included into this group, which means you can also invite the author of the book to join the course or class or gathering.

2- Collectively send out announcements
   Take full use of the announcement section in Google Classroom and send out announcements and resources to the entire group instead of sending in emails. This Google Classroom can serve as a reservoir for memos, documents, photos and so on even if it isn't used as a 'class' to teach.

3- Presentations, workshops and training
Anyone wanting to share knowledge can open a Google Classroom and invite audiences to join their classes on a specific event. Want to give your YouTube exclusive excess on how to make your famous devil food cake? Open a Google Classroom! Want to give your lucky subscribers access on a 30 days fitness bootcamp? Open a Google Classroom! Want to conduct a workshop on the best ways on using social media? Open a Google Classroom. Have a closed presentation to a select committee? Get them to join your Google Classroom!

4- Digital Cert Awards
Give out digital certs for assignments that your students have completed as is shows that you recognize their accomplishment. Badges can be given out for completing a variety of tasks whether it is an assignment or completion of a course or workshop, anything to make the students feel accomplished.

5- Create Online Courses
If you are an expert in your field, promote your classes to your audience. Get them to

join in at a select time and date for your audience to register and you can conduct your classes through slideshows, documents, giving an introduction video and so on. Students will still be able to access the information given even after the class is over.

6- Tutoring

Are you a teaching assistant or a tutor? Give your students access to your Google Classroom when you are not able to meet them physically for a class. Ensure that they have their assignments and the due date to submit. You can also connect all your students into one Google Classroom and give them tips, advice and exercises after their core subject classes.

7- PTA Meetings

Many of your teachers would have had this problem where it is hard to find the right time and date to get majority of parents and teachers available for a PTA meeting. Unless if it is a necessary meeting that requires voting and things like that, certain PTA meetings can be conducted online, at the convenience of the class teacher and all the parents in that class.

Any information shared in this meeting is private & confidential.

8- Teacher-Parent Group
   Don't want to go through the hassle of creating a group message on Whatsapp or iMessage? Don't feel like you want to share your private phone number with Parents? Then create a group in your Google Classroom. Give parents updates on activities in the school which require their participation. You can also private message certain parents to discuss a student's performance and find ways that can help the student cope.

Conducting training or workshop or a class is much easier when there is a program specifically built for this purpose. Although the education sector was the primary focus group for Google Classroom, it has now become a tool that can be used by millions of people wanting to share whatever knowledge they have in the simplest way possible using the easiest platform.

Conclusion

Thank for making it through to the end of Google Classroom (Technical), *A Simple, Concise & Complete Guide to Take Your Classroom Digital.*

Let's hope it was informative and able to provide you with all of the tools you need to achieve your goals of making your classroom digital.

We hope that you gained valuable insight into the world of Google Classroom.
One of the best things about Google Classroom is that is it fuss free, especially now that you only need a Gmail account to create a classroom.

Setting up a classroom is easy and it does not increase a teacher's headache or stress trying to figure out how to use this application.

Invites to collaborate, share information, give announcements and assignments are done easily and quickly.

Students also get immense benefits from Google Classroom as they can seek advice, get updates, connect to a teacher and collaborate with their friends on assignments.

Google Classroom is a simple and paperless solution for anyone wanting to start a class but do not want to go through the hassle of printing and distributing information. This makes managing students work easier and in a much more efficient manner, all in one place, with a few clicks.

Combined with other Google products such as Google Maps, Calendar and Google Docs, Google Classroom is definitely an avenue that minimizes excuses of having no time, not enough resources or the ultimate 'my dog ate my homework' scenario.

If you found this book useful, then get your hands on the second part of Google Classroom that looks at '50 Ways You Can Use Google Classroom to Effectively Implement Digital Tools in Your Classroom'

Finally, if you found this book useful in any way, a review on Amazon is always appreciated!

www.ingramcontent.com/pod-product-compliance
Lightning Source LLC
Chambersburg PA
CBHW070304230526
45470CB00002B/722